KANGAROOS

KANGAROOS

ODYSSEYS

MELISSA GISH

CREATIVE EDUCATION · CREATIVE PAPERBACKS

Published by Creative Education and Creative Paperbacks
P.O. Box 227, Mankato, Minnesota 56002
Creative Education and Creative Paperbacks
are imprints of The Creative Company
www.thecreativecompany.us

Design by Blue Design (www.bluedes.com)
Art direction by Wyeth Morgan

Images by Dreamstime/Martin Pelanek, cover, 2; flickr/Biodiversity Heritage Library, 38; Getty Images/AzmanL, 70–71, Grant Faint, 50, Jami Tarris, 6, 11, 57, 61, 62–63, Joao Inacio, 27, Lea Scaddan, 4–5, 8, 58, MARK RALSTON, 40, Martin Harvey, 17, Matt Deakin, 24, Raimund Linke, 49, Robbie Goodall, 22; Pexels/Ethan Brooke, 42–43, 75, Frank Schrader, 32–33; Unsplash/Simone Viani, 64; Wikimedia Commons/Charles J. Sharp, 12, H. J. Richardson, 54, Kürschner, 46, Peter Purves Smith, 69, Thennicke, 28, Tim Williams, 37, VmeganV, 18

Library of Congress Cataloging-in-Publication Data
Library of Congress Cataloging-in-Publication Data
Names: Gish, Melissa, author.
Title: Kangaroos / Melissa Gish.
Other titles: Kangaroos (Odysseys in the wild)
Description: Mankato, Minnesota : Creative Education and Creative Paperbacks, [2026] | Series: Odysseys in the wild | Includes bibliographical references and index. | Audience: Ages 12–15 | Audience: Grades 7–9 | Summary: "Discover all there is to know about kangaroos, the marvelous mammals from Down Under, in this high-school nonfiction text exploring the lives of these wild animals"– Provided by publisher.
Identifiers: LCCN 2024045519 (print) | LCCN 2024045520 (ebook) | ISBN 9798889896579 (library binding) | ISBN 9781682778234 (paperback) | ISBN 9798889897378 (ebook)
Subjects: LCSH: Kangaroos–Juvenile literature.
Classification: LCC QL737.M35 G576 2026 (print) | LCC QL737.M35 (ebook) | DDC 599.2/22–dc23/eng/20241219
LC record available at https://lccn.loc.gov/2024045519
LC ebook record available at https://lccn.loc.gov/2024045520

Printed in India

Mother kangaroos are highly protective of their young.

CONTENTS

Introduction

Like an orange ball floating on a sea of grass, the sun touches the horizon. Here in Australia's Warrumbungle National Park, kangaroos stand up and stretch after a hot, lazy afternoon spent sleeping in the shade of scattered trees. As the temperature drops, the kangaroos begin their evening ritual of munching on grass and tender shrubs.

OPPOSITE: Eastern grey kangaroos are the most common mammals found in Warrumbungle National Park, Australia.

female rises, and her offspring, called a joey, peeks its head out of the pouch on its mother's belly. It stretches out its arms, leans down, and touches the grass with its tiny paws. Then it slips out of the pouch for the first time in its life, hopping in small circles close to its mother, bending to nibble the slender grass, and then leaning back on its tail to practice balancing. Suddenly, a hawk screeches overhead, startling the joey, who quickly leaps headfirst back into its mother's pouch. It will explore the world again tomorrow.

Kangaroo joey in its mother's pouch

Meet the Kangaroo

About 54 members of the *Macropodidae* family—including kangaroos—live in the open spaces of Australia, New Zealand, New Guinea, and some surrounding islands. Macropods are characterized by their straight cutting teeth, powerful tails, and large hind feet.

OPPOSITE: Kangaroos have adapted to various habitats in Australia and the surrounding islands, from dry, warm plains to more temperate forests.

The family includes the threatened quokka, along with various wallabies, pademelons, and tree kangaroos. It also includes six species of animals that have hind feet measuring 10 inches (25 centimeters) or longer, leading scientists to commonly group them together as kangaroos.

These six species of kangaroos are the eastern grey, western grey, red, and antilopine kangaroos and the black and common wallaroos. The word "macropod" means "large foot" in Latin and Greek, and all macropods, except for tree kangaroos, have large hind feet and long hind legs. The word "kangaroo" is a variation of *gungurru*, a northeast Australian **Aboriginal** word for a kind of kangaroo that is now extinct.

Kangaroos are mammals. All mammals produce milk to feed their young and, except for the egg-laying platypuses and echidnas, give birth to live offspring. Mammals are also warm-blooded. This means their bodies try to maintain a healthy, constant temperature that is usually warmer than their surroundings. Kangaroos live in hot climates and cool themselves by licking their front limbs. The saliva **evaporates** to cool the blood just beneath the skin. Kangaroos also pant like dogs, which reduces their internal body temperature.

Kangaroos make up only a fraction of the nearly 300 **marsupial** species on Earth. The word "marsupial" means "purse bearer." Female marsupials have thick folds of flesh on their abdomens where they carry their young. Kangaroos are the world's largest marsupials. They share their **Australasian** habitat with smaller marsupials such as koalas, wombats, Tasmanian devils, and banded anteaters.

Male kangaroos are called boomers. Females are called flyers. The red kangaroo is the largest of the kangaroo species. Males weigh up to 200 pounds (91 kilograms) and stand 6.6 feet (2 meters) tall. Their body length averages 5 feet (1.5 m), and their tails can add up to 4 more feet (1.2 m). Females are smaller, rarely

Male and female red kangaroos

Is It a Kangaroo?

Wallabies are another famous animal from the "Land Down Under" with distinguishing features such as stomach pouches and the ability to hop great distances. This leads them to be easily mistaken for kangaroos. While wallabies and kangaroos are both macropods and share many similarities, upon closer look they have a few key differences. Wallabies are much smaller than kangaroos, growing to only about 3 feet (0.9 m) while kangaroos can tower at more than 6 feet (1.8 m) tall. Wallabies also have shorter legs, and their fur is much more colorful, usually combining two or more colors, while kangaroo fur remains a single shade of brown.

reaching 6 feet in height (1.8 m). The other kangaroo species are nearly as large as the red kangaroo but have adopted varying characteristics to suit their environments.

Kangaroo fur, called pelage, is thick and velvety, protecting kangaroos from thorns and brambles and insulating the animals from the heat of their environment. Grey kangaroos are named for their primary coloring. They also have whitish underparts, black tips on their tails, and fine hair on their jaw, mouth, and nose—the part of an animal's face called a muzzle. The male red kangaroo is a reddish-brown color, while the female is bluish gray. Red kangaroos have fine hair only on their muzzles. Wallaroos, reddish in color with whitish underparts, have no hair on their muzzles.

All kangaroos are herbivores, meaning they eat only vegetation such as grass, shrubs, and the leaves of small

"ALL KANGAROOS ARE HERBIVORES, MEANING THEY EAT ONLY VEGETATION SUCH AS GRASS, SHRUBS, AND THE LEAVES OF SMALL TREES."

trees. A kangaroo's front teeth are designed for clipping grass close to the ground. The back teeth, called molars, are sharply ridged for grinding food into pulp. Kangaroos have four pairs of molars in each jaw. Constant chewing causes the front molars to wear down to the roots and eventually fall out. When this happens, the back molars move forward. By the time a kangaroo is very old—about

18 to 20 years—it may have only one upper and one lower molar left in its mouth.

An adult kangaroo eats about 15 pounds (6.8 kg) of vegetation each day. Digesting so much plant fiber requires a special stomach—one with four chambers, or sections. Food passes through the first chamber, called the rumen, where bacteria and acids soften it. Then the food is regurgitated, or brought back up to the mouth. This food mass, called a cud, is chewed again. After it is swallowed, the cud is then fully digested. Kangaroos can survive for months without drinking water, as the vegetation they eat provides plenty of moisture. If plants dry out and kangaroos need water, they will dig pits about 3 feet (0.9 m) deep until they reach water.

Red kangaroos live in central Australia's hot, arid plains, an area known as the "Red Center." Grey and

OPPOSITE Kangaroos spend large portions of the day sleeping and are most active at dawn and dusk.

antilopine kangaroos cannot tolerate that much heat, so they live most everywhere else in Australia, where the land is more forested. Wallaroos can be found on the rocky landscapes and steep slopes of far northern and northeastern Australia. They rest in caves during the hottest parts of the day.

Because wallaroos rarely travel long distances to forage, their bodies adapted by developing shorter, stockier legs than grey and red kangaroos, which may travel more than 100 miles (161 kilometers) to find

The Five-Legged Animal

Most mammals have tails that serve various purposes, but kangaroo tails are special. Studies show that a kangaroo's tail functions like a fifth leg. When kangaroos hop from one place to the next, their tails provide support and power to propel them forward, similar to human legs when walking or running. Some research even suggests that a kangaroo's tail alone provides more power than its front and back legs combined. Kangaroos use their tails to travel at fast speeds for long periods of time without getting tired. Some species of kangaroos can momentarily balance their entire bodies on their tails.

food. To travel, kangaroos do not run; they jump. The kangaroo's body, with its powerful hind legs and large hind feet, is designed for jumping. About 75 percent of a kangaroo's body weight is in its hindquarters, which are very muscular.

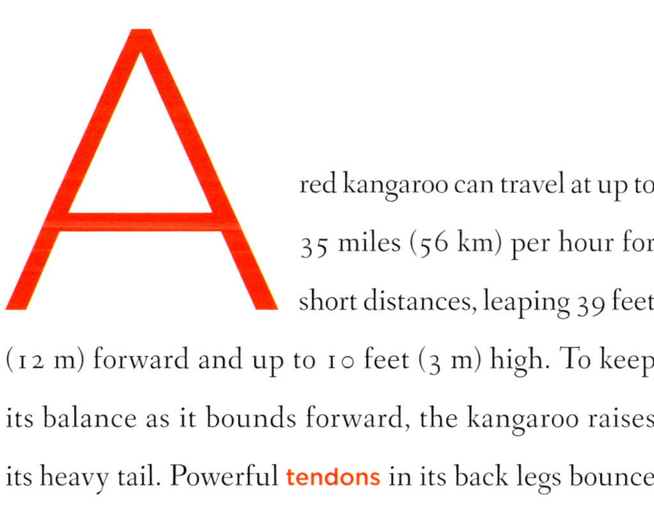

A red kangaroo can travel at up to 35 miles (56 km) per hour for short distances, leaping 39 feet (12 m) forward and up to 10 feet (3 m) high. To keep its balance as it bounds forward, the kangaroo raises its heavy tail. Powerful **tendons** in its back legs bounce

"TO KEEP ITS BALANCE AS IT BOUNDS FORWARD, THE KANGAROO RAISES ITS HEAVY TAIL."

like springs to allow a kangaroo to hop steadily for hours without tiring. The average speed of a cruising kangaroo is 20 miles (32 km) per hour.

Kangaroos also shuffle along slowly by leaning on their front paws and swinging their back legs up together from underneath them. This is called crawl-walking. The kangaroo has four toes on each hind foot.

The two inner toes are partially fused and share a long, sharp claw that is used for grooming and defense. The kangaroo's front paws have five "fingers" with sharp claws that grip food and are also used for grooming. Kangaroos spend a lot of time grooming to keep pesky insects out of their fur.

Kangaroo front and back paws

Mob Life

Kangaroos are social animals, living in groups called mobs that consist of at least 2 or 3 individuals and may have up to 100. A dominant male leads the mob, which includes several adult females, as well as both male and female juveniles. Every kangaroo knows its place in the mob's **hierarchy**.

OPPOSITE: Kangaroo mobs are also called troops, herds, or courts.

The head female, usually the oldest and the one charged with teaching younger kangaroos how to get along in the mob, is called the matriarch. A sentry, or lookout, is responsible for staying alert to potential threats and for warning the rest of the mob when danger approaches. Sentries will sound the alarm by thumping the ground with their feet. The mob will then scatter, with only mothers and their joeys staying together.

Because adult male kangaroos are large animals, they have no natural predators. However, joeys may be snatched up by large birds of prey such as wedge-tailed eagles. Joeys and small females are also preyed upon by packs of wild dogs called dingoes. The lead dingo will often chase the kangaroo toward its pack mates, which hide and wait. Then they leap on the kangaroo and bite its neck.

Kangaroos defend themselves with clawed swipes and kicks that can tear a dingo apart. Kangaroos may also allow dingoes to chase them into a body of water. There a kangaroo will grab the dingo in its front paws and hold it underwater until it drowns. However, if a pack works together to attack a kangaroo, the dogs are often successful in bringing it down.

Male kangaroos kick and punch for a different reason. Kangaroos playfight to show each other their skill and strength in a non-threatening way, but when the time comes for mating, males fight seriously

Kangaroos are smarter and easier to train than dogs, but they are also more independent and difficult to control.

to win a female. Fighting kangaroos swipe and jab with their forearms to knock their opponents off balance. They may also lean back on their tails and raise their rear legs up to deliver powerful kicks. When one kangaroo has had enough, he will retreat, leaving the winner with his choice of females.

Female kangaroos are choosy about potential mates. When approached, a female may swat many suitors away before allowing one of the right size and healthy appearance to court her. The female stands still as the male begins making clicking sounds close to her head. Then he will stroke her chest, neck, and tail. If she is displeased, the female will hop away, but if she likes the male, she will allow him to mate with her.

Kangaroos have no specific mating season and breed year round. Females have one joey at a time. After a gestation of 29 to 38 days, a joey the length and weight of a single paper clip is born. It is pink, hairless, and blind, yet it already has clawed front paws and a good sense of smell, which it uses immediately. Clutching its mother's fur and following its nose, the joey takes about three minutes to drag itself from its mother's birth canal to the inside of her pouch. There it clamps its mouth around one of her four teats, and the mother pumps milk down the joey's throat. The joey nurses for about three months, growing very slowly. Eventually, the mother's milk changes to include more protein and fat. This makes the joey grow more quickly.

Wallaroo and red kangaroo joeys are old enough to leave the pouch at about six months of age. Grey kangaroo joeys spend an additional two months in the pouch. Once they leave the pouch to begin feeding on grass, they can still easily jump back into the pouch for protection and to continue feeding on their mother's milk, which they will need for an additional six to eight months. Wallaroos and red kangaroos are independent of their mothers at 12 months old, grey kangaroos at 18 months. By the time female red and grey kangaroos are 14 to 18 months old, they are old enough to mate and have joeys of their own.

In all species of kangaroos, except for the western grey, reproduction is tied to the climate in which the kangaroos live. During times of drought,

Jumpers and Climbers

Tree kangaroos are close relatives of the kangaroo. They are much smaller, have longer tails, and look more like lemurs than kangaroos. They are also the only macropods that climb and live in trees. There are 13 species of tree kangaroos, all of which are found in the rainforests of New Guinea, Indonesia, and northeastern Australia. Several tree kangaroo species are in danger of extinction due to habitat loss and hunting. These include the golden-mantled and Wondiwoi tree kangaroos. Approximately 50 Wondiwoi tree kangaroos are left in the wild.

when vegetation is scarce, a female kangaroo that becomes pregnant while still carrying a joey in her pouch may put the new one "on hold" until her current joey leaves the pouch. During this period, called embryonic diapause, the **embryo** stops developing for up to 11 months until environmental conditions improve. This suspension of birth ensures that the new joey will have a food source when it is ready to eat grass. When the second joey is born and crawls into its mother's pouch, it is fed the kind of milk that newborns need, while the first joey, returning to the pouch only to nurse, is fed the kind of milk that older joeys need.

hen food is in short supply, kangaroos cannot spend their days lounging in the shade and their nights nibbling on grass. Instead, they must constantly search for food, and this often leads them to cities and towns. Places that are watered—parks, lawns, golf courses, and sports fields—are places where kangaroos will gather. When such places

Boxing Kangaroos

The boxing kangaroo is a famous national symbol in Australia. It features a yellow kangaroo with red boxing gloves on a green backdrop. Meant to represent the fighting spirit of the country's citizens, the boxing kangaroo was painted on the sides of Australian military vehicles during World War II, it has been featured on flags and postage stamps, and it is also the symbol for the Australian Olympic Team. Kangaroos sometimes appear to be "boxing" when they fight each other for dominance. There have also been kangaroos that were trained to fight in real boxing matches against humans, but this practice has been largely discontinued.

are the only green environments available, kangaroos will refuse to be chased away.

Over a normal kangaroo's life span of 12 to 18 years in the wild (which can be shorter or longer, depending on food supply) or up to 25 years in captivity, a kangaroo may become accustomed to sharing its environment with people. Kangaroos may appear tame around people, but they are wild animals that can grow to be as big and as heavy as an adult man, making them dangerous. People have been swatted, clawed, and kicked by kangaroos that are behaving as they would against predators in the wild. Even pets can come into deadly contact with kangaroos, which view dogs as threats.

Kangaroos have excellent hearing and turn their ears to pick up sounds.

Kangaroos and People

Kangaroos have been a major part of Aboriginal culture since the first people arrived on the continent from Asia about 45,000 years ago. The native peoples had formed about 700 groups that spoke more than 200 different languages by the time the first Europeans visited Australia in the late 1700s.

The early Aboriginal people, who had more than 40 different names for the kangaroo, lived as hunter-gatherers. The women collected fruits, berries, plants, and bird eggs, and the men hunted large, flightless birds called emus, geese, and a variety of mammals as food sources. Small kangaroo relatives, including wallabies and bettongs, were caught with traps made of grass rope. Kangaroos were hunted with boomerangs, which are also called throw sticks. A boomerang is a flat wooden weapon carved like a bent stick. A skilled boomerang thrower can knock a kangaroo—even a large boomer—off its feet. "Boomerang" is an Aboriginal word.

Locations where kangaroos were regularly hunted were called *kangaloola*. None of the kangaroo was wasted. Its meat was either cooked or dried in the sun to make jerky. Its bones were made into tools, weapons,

OPPOSITE Coats and other clothing have been made from kangaroo skins from the time of the early Aboriginal people to present day.

and sewing needles. The tendons from its tail were used as thread to sew clothing made from the kangaroo's hide and the skins of other animals. Kangaroo hides were also used as blankets and rugs, and they were made into bags for storing water.

The native peoples of Australia also included the kangaroo in their storytelling and religion. The early Aboriginal people had no written language. Instead, they used images to document stories. They painted and stenciled images of people and animals on rock faces and cave walls using charcoal, white clay, and ochre. Yellow and red ochre are **pigments** made from clay. Across Australia, more than 100,000 Aboriginal rock art sites have been discovered.

The oldest kangaroo paintings ever discovered exist in Kakadu National Park, east of the seaside city of Darwin in Australia's Northern Territory. The cave paintings in Kakadu are more than 10,000 years old. About 100 miles (161 km) from the city of Sydney, the capital of New South Wales, in the far southeastern corner of Australia,

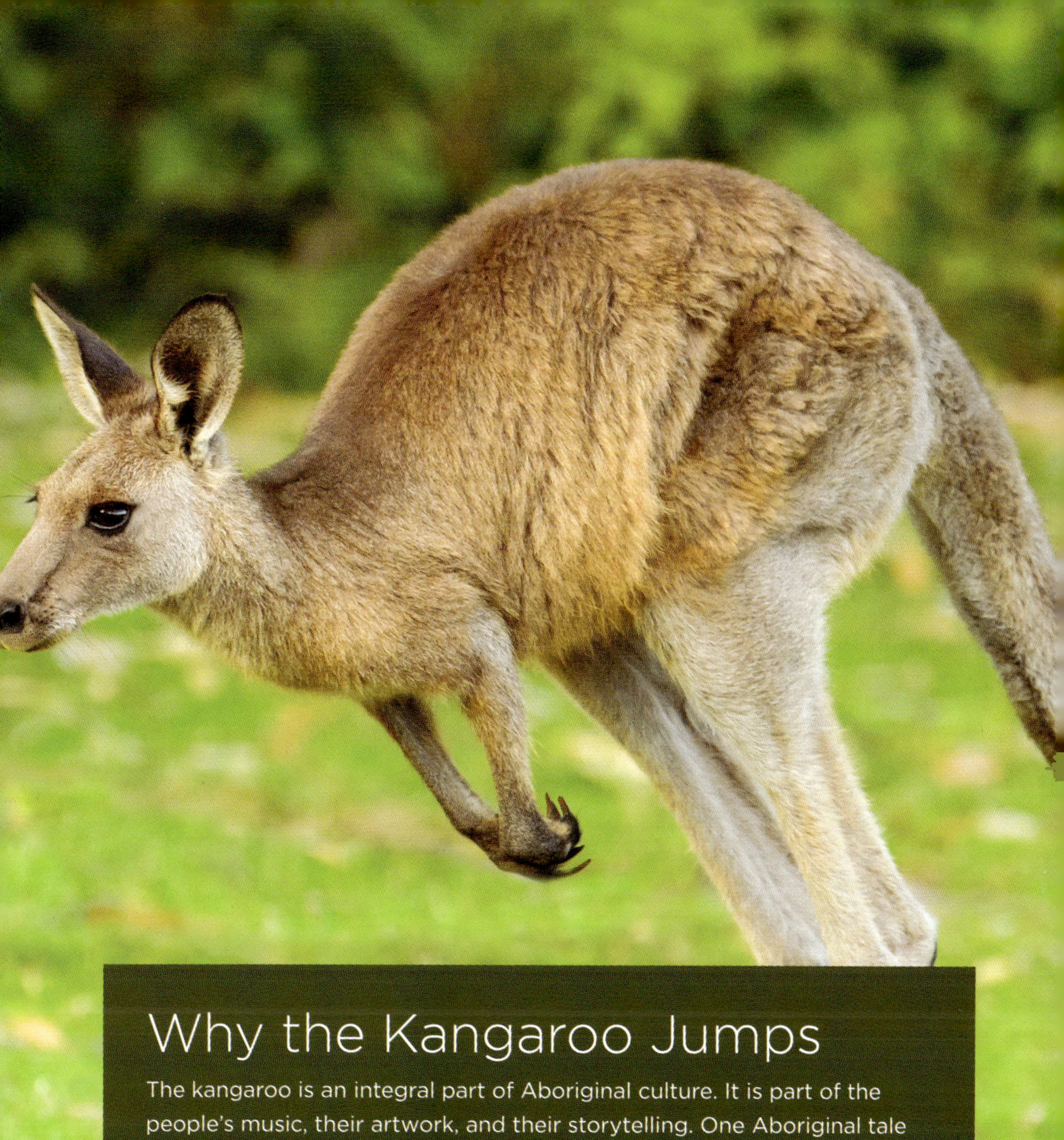

Why the Kangaroo Jumps

The kangaroo is an integral part of Aboriginal culture. It is part of the people's music, their artwork, and their storytelling. One Aboriginal tale explains why the kangaroo jumps instead of runs. The tale speaks of a kangaroo in the forest that was being hunted by a human. Unable to outrun the man on its four legs, the kangaroo observed how the man ran and began to hop using only its two hind legs. The kangaroo was able to move faster this way, and it eventually escaped the hunter. After, it showed the other kangaroos in the forest how to jump. From then on, all kangaroos moved this way.

About 400 different Aboriginal groups exist in Australia.

a 4,000-year-old Aboriginal rock art site was discovered in 2003. Not far from that site, more than 200 individual images—including those of lizards, birds, wombats, and kangaroos—were found in a recently explored area of Wollemi National Park. Also painted on the rock faces are figures called therianthropes (*thee-ree-ANTH-ropes*). These are half-human, half-animal figures. Some of the therianthropes show humans with bird heads, and some show human-like kangaroos.

These figures were tied into the many religious beliefs about animals that were held by the Aboriginal people. Each group had a **totem** that symbolized its connection to its ancestors and to the land. The Aboriginal people believed that the spirits of their ancestors lived in the totem animals. Some totems were plants, and some were animals. Emus were common totems, as were kangaroos.

If a man identified with kangaroos, he was called a kangaroo man. In most groups, the rules regarding totems said that a kangaroo man could not eat kangaroo meat that was bloody. It had to be thoroughly cooked before he could eat it. In some kangaroo groups, killing kangaroos was forbidden.

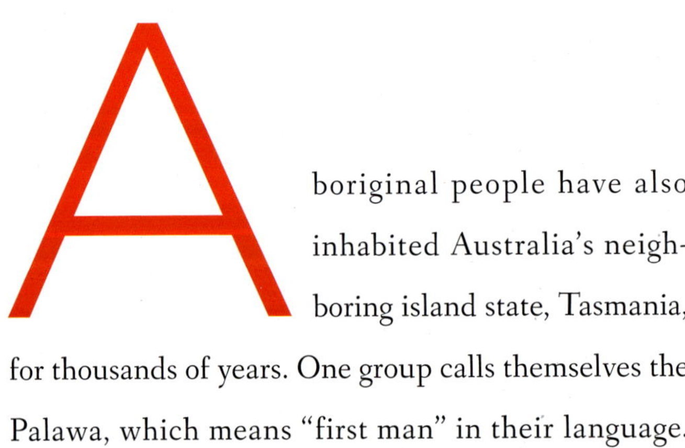

Aboriginal people have also inhabited Australia's neighboring island state, Tasmania, for thousands of years. One group calls themselves the Palawa, which means "first man" in their language.

They believe that a creation spirit made humans from a kangaroo. Long ago, people wore no clothing except a kangaroo skin draped over their shoulders like a poncho, and they painted their bodies with a mixture of ochre and bird fat. Even though the Palawa people hunted kangaroos and wallabies, they respected these animals, showing their appreciation for a killed animal's sacrifice with prayers and dances.

A popular myth about the kangaroo's discovery by 18th-century Europeans tells that when British captain James Cook visited Australia in 1770, he pointed to a kangaroo and asked an Aboriginal person what it was. The person supposedly said, "Kangaroo," which supposedly meant "I don't know" in his language. While this story has been proven false, the myth persists, and many people still believe that kangaroos are named for

OPPOSITE Kangaroo populations in Australia started to decrease once European settlers arrived in the 1800s.

the phrase “I don’t know.” In truth, Captain Cook did visit Australia in 1770 after his ship, the *Endeavour*, ran aground on a coral reef and needed repairs before returning to sea. Aboard Cook’s ship was the naturalist Joseph Banks, who recorded his impressions of the kangaroos he saw and learned the true Aboriginal name for the animal—*gungurru*.

Europeans quickly discovered the commercial value of kangaroos and began hunting them for their skins and meat. Although their numbers dropped drastically in the 1800s, kangaroos have always steadily reproduced and have, over the generations, been able to maintain a healthy population across most parts of Australia. As white settlements grew and cities developed in the 19th century, the kangaroo became a part of the art and architecture. Kangaroo carvings appeared atop buildings, images of kangaroos adorned windows and doors, and kangaroo statues were shaped out of stone and cast out of metal to decorate parks and city centers.

People's regard for the kangaroo as a symbol of Australia's unique wildlife remains strong today. Australia's **coat of arms** includes a kangaroo and an emu, symbols of the nation's natural heritage. The animals also represent progression toward the future because kangaroos and emus, unable to move backward, always move forward. Leaping kangaroos—the enduring symbol of Australia's positive attitude—serve as the emblems of Australia's military branches, the Qantas airline, and several Australian coins. In 2022, the Royal Australian Government released a new set of silver and gold coins as part of their Impressions of Australia series. On the front of the coins was an image of Queen Elizabeth II of England, then

The red kangaroo remains a famous symbol for Australia.

ruler of Australia. On the back, a kangaroo stood on rocks next to a pair of trees with images of grain and a sun in the background. According to the mint, the coins "immortalized Australia's most iconic animal."

Research and Conservation

The oldest known kangaroo ancestor was the size of a chipmunk. *Sinodelphys szalayi* was a marsupial that lived in China about 125 million years ago. It ran on all fours and ate insects and worms like modern opossums.

OPPOSITE: All baby marsupials are called joeys, whether they're kangaroos (pictured), koalas, wombats, or other marsupial species.

About 25 million years ago, as kangaroo ancestors grew larger, they developed fangs and may have climbed trees. The first true kangaroos—hopping animals that resemble modern kangaroos—emerged in Australia about 15 million years ago. For millions of years, the giant short-faced kangaroo (*Procoptodon goliah*) dominated Australia's plains and forests. It stood 10 feet (3 m) tall and had massive claws and sharp hooves. It existed until just about 40,000 years ago, when a major drought wiped out most of Australia's large animals and birds. Kangaroos grew smaller to survive and eventually became the animals we know today.

Kangaroos are legally protected in all territories of Australia, but for decades, these animals have been the subject of much disagreement among lawmakers, citizens, and conservationists. Some people see kangaroos

The Ripped 'Roo

One kangaroo that rose to stardom in the early 21st century was a red kangaroo named Roger, also known as Roger the "ripped" or "buff" kangaroo. Roger became an internet sensation for his abnormally muscled body, which made him look more superhuman than animal. After being orphaned at a young age, Roger was taken in by The Kangaroo Sanctuary in Alice Springs, Australia. He grew to be 6 feet, 7 inches (2 m) tall, and he was known as a fierce yet lovable fighter. Roger was featured in the BBC documentary series *Kangaroo Dundee* alongside his caretaker, Chris "Brolga" Barnes. He died in 2018 of natural causes, and a monument dedicated to his memory stands in the sanctuary.

When they fight, male kangaroos risk scarring, broken bones, and internal injuries that may lead to death.

Kangaroos and livestock often compete for valuable resources such as food.

as a nuisance animal that must be controlled. Others see Australia's national symbol as a valuable part of the ecosystem that must be preserved. Still others appreciate kangaroos for their commercial value. Scientific study and wildlife management programs function to create a balance between the need to control kangaroo populations and the need to protect the animals from senseless slaughter.

Many sheep farmers consider kangaroos a nuisance because their animals must compete with kangaroos for grazing grass. A study conducted by the University of New South Wales in Sydney revealed that kangaroos have no negative effects on sheep farms—yet sheep farmers are legally allowed to shoot kangaroos that come onto their land. In urban areas, kangaroos can pose a danger to people. Since 1990, the number of kangaroos in Australia's capital city of Canberra has grown exponentially, and more than 500 kangaroo-caused automobile accidents are reported each year. Other cities across Australia must also contend with ever-growing populations of kangaroos.

A method of kangaroo population control that provides immediate results is culling, or killing off a portion of the population. Australia's citizens are divided on the issue, though. Some people, such as farmers and ranchers, want to see kangaroos destroyed, while others want the government to come up with a better way of dealing with the overpopulation. Wildlife conservationists have filed legal complaints against the Australian government, calling kangaroo culling cruel and unnecessary. Plans to cull large numbers of kangaroos within cities are often halted by public protest, yet most urban areas in Australia

“MANY COMPANIES IN AUSTRALIA ARE ALLOWED TO SHOOT KANGAROOS AND SELL THEIR HIDES AND MEAT.”

still regularly (and discreetly) remove kangaroos—either by trapping and relocating them or by shooting them.

Another method employed to decrease kangaroo populations involves the commercial wildlife trade. Many companies in Australia are allowed to shoot kangaroos and sell their hides and meat. The number of kangaroos that can be "harvested" depends each year on the number of kangaroos counted by kangaroo management groups

OPPOSITE The practice of kangaroo culling dates back to the late 1800s.

funded by the government. Since 1978, various groups have flown small airplanes over kangaroo habitats and counted the kangaroos in given areas. Then a percentage of that total count is declared to be the annual quota, or maximum number of kangaroos that can be killed for commercial purposes that year.

Over the past 25 years, populations of kangaroos have bounced between 15 and 50 million. During times of fair weather and plenty of rainfall, kangaroos are abundant, and culling quotas are set high. In times of drought, kangaroo populations may plummet, and quotas are set low. The quotas are set to cull about 10 to 20 percent of the population. Most years, the cull targets are not met. The annual quotas have become more **controversial** over the years. In 2022, the Government of the Australian Capital Territory (ACT) invested $1.2 million in a **contraception**

‘Kangaroos Are Not Shoes’

Kangaroo leather is used in products such as belts, gloves, wallets, and, most controversially, soccer cleats. Top soccer-cleat manufacturers have historically favored kangaroo leather because it is lightweight and durable. In 2020, people started becoming more concerned about the number of kangaroos being killed to produce soccer cleats. One campaign called “Kangaroos Are Not Shoes” organized protests, fundraisers, and petitions to ban the use of kangaroo leather. As of 2024, Nike, Puma, and New Balance had all pledged to stop using kangaroo leather in their products. Many professional soccer players also take a stand by choosing to wear cleats made from synthetic materials, such as nylon.

program as a more humane population management option. The program began in Canberra's parks, where rangers sedated female kangaroos with darts, injected them with a long-lasting fertility control vaccine, and fitted them with ear tags for monitoring. While this method will reduce culling numbers in the future, the program is suitable for managing only small populations.

The animals that are harvested during the cull support an industry that generates more than $200 million per year. Australia exports kangaroo meat and skins to more than 55 nations around the world, including the United States. Some researchers have suggested that kangaroo meat is more environmentally friendly than beef. Kangaroo skins are used in leather products, including tennis shoes, golf gloves, and baseball mitts. Kangaroo leather products are labeled as "K leather" or "RKT," which stands for "rubberized kangaroo technology."

Another field in which kangaroos have been useful is human medical research. The Kangaroo Genome Project is run by the ARC Centre of Excellence for Kangaroo Genomics, or KanGO, in Australia. Several Australian universities and research organizations make up KanGO, which researches kangaroo **genetics**. It mapped the complete kangaroo genome in 2007. Researchers have used the map to study how joeys develop in a pouch rather than a uterus, like other mammals, and, in turn, to learn more about human development.

KanGO is also examining the possibility of duplicating kangaroo genes and inserting them into dairy cows to help them produce more milk, since kangaroos produce highly rich milk for their growing offspring. Kangaroo milk also contains powerful **antibiotics** that protect joeys from illness. Scientists are working toward an understanding of these substances with the goal of duplicating them for human use.

Kangaroos are cherished symbols of Australia, the unique "Land Down Under." Apart from providing endless entertainment for tourists, kangaroos serve many other valuable roles in human society. Proper management of kangaroo populations will ensure that Australia's boomers and flyers will always have a special land to call home.

Left-Pawed

What bounces everywhere, has a stomach pouch, and is left-handed? A 2015 study conducted out of Saint Petersburg State University in Russia found that most red and eastern grey kangaroos are left hand dominant. Whereas only an estimated 10 percent of humans are left-handed, these kangaroos may use their left hands up to 95 percent of the time, specifically when performing precision-based tasks such as eating and grooming. Before this study, being dominant in one hand over the other was believed to be a trait shared only by primates, such as humans. Scientists are eager to discover what other similarities kangaroos and humans might have.

Selected Bibliography

ACT Government. "Kangaroos." https://www.environment.act.gov.au/parksconservation/plants-and-animals/urbanwildlife/kangaroos.

Dickman, Christopher. *A Fragile Balance: The Extraordinary Story of Australian Marsupials.* Chicago: The University of Chicago Press, 2007.

The Editors of Encyclopedia Britannica. "Kangaroo." Encyclopedia Britannica. https://www.britannica.com/animal/kangaroo.

Flannery, Tim. *Chasing Kangaroos: A Continent, a Scientist, and a Search for the World's Most Extraordinary Creature.* New York: Grove Press, 2007.

Kangaroo Industry Association of Australia. "About KIAA." https://www.kangarooindustry.com/en/about-us/.

Pickrell, John. *Flames of Extinction: The Race to Save Australia's Threatened Wildlife.* Sydney, New South Wales, Australia: NewSouth, 2021.

Glossary

Aboriginal of or relating to the people who inhabited Australia before the arrival of European settlers

adapt to change to improve the chances of survival in an environment

antibiotic a medicine that kills or disables the growth of bacteria, which are living organisms that cannot be seen except under a microscope

Australasian of Australia, New Zealand, or the islands northeast of Australia

coat of arms the official symbol of a family, state, nation, or other group

commercial suitable for business and to gain a profit rather than for personal reasons

contraception a method of preventing pregnancy

controversial relating to or causing much disagreement or argument

ecosystem a community of organisms that live together in an environment

embryo an unborn or unhatched offspring in its early stages of development

evaporate to change from liquid to invisible vapor or gas

genetics relating to genes, the basic physical units of heredity

hierarchy a system in which people, animals, or things are ranked in importance one above another

marsupial a mammal whose young are born early and further develop in a pouch on the mother's body

nuisance something annoying or harmful to people or the land

pigment a material or substance present in the tissues of animals or plants that gives them their natural coloring

stencil to paint or mark around a sheet of material that has been cut in a certain shape

tendon a tough, inelastic tissue that connects a muscle to bone

totem an object, animal, or plant respected as a symbol of a group and often used in ceremonies and rituals

Websites

Fun Facts about Kangaroos

https://www.australia.com/en-us/things-to-do/wildlife/kangaroo-interesting-facts.html

Discover more interesting facts about kangaroos.

Kangaroo Boxing Fight

https://www.youtube.com/watch?v=WCcLMNcWZOc

Watch kangaroos bounce and kick as they practice fighting.

Red Kangaroo

https://australian.museum/learn/animals/mammals/red-kangaroo/

Learn more about Australia's iconic red kangaroo.

Index